Explore the Tour of Mont Blanc

Gareth McCormack

Rucksack Readers

Explore the Tour of Mont Blanc: a Rucksack Reader

First published in 2005 by Rucksack Readers, Landrick Lodge, Dunblane, FK15 0HY, UK

Telephone   +44/0 1786 824 696
Website   **www.rucsacs.com**
Email   info@rucsacs.com

Distributed in North America by Interlink Publishing, 46 Crosby Street, Northampton, Mass, 01060, USA (www.interlinkbooks.com)

© text, design and layout copyright 2005 Rucksack Readers; photographs © Gareth McCormack.

Gareth McCormack reserves the right to be identified as the author of this work.

All rights reserved. No part of this publication may be reproduced, stored in a retrieval system, or transmitted in any form or by any means (electronic, mechanical, photocopying, recording or otherwise) without prior permission in writing from the publisher.

ISBN 1-898481-20-2

British Library Cataloguing in Publication Data: a catalogue record for this book is available from the British Library.

Designed in Scotland by **WorkHorse** (info@workhorse.co.uk)

Colour separation by HK Scanner Arts International Ltd in Hong Kong

Printed in China by Hong Kong Graphics & Printing Ltd

The maps in this book were created for the purpose by The XYZ Digital Map Company © 2005.

Publisher's note

Walking in the Alps can involve possible hazards which have been explained by the author. However, individuals are responsible for their own welfare and safety, and the publisher cannot accept responsibility for any ill-health or injury to readers of this book, however caused.

All information has been checked carefully prior to publication, but change is inevitable. We are always delighted to hear from any reader who can provide an update or any feedback on the walk or the book, ideally by email to **info@rucsacs.com.**

Explore the Tour of Mont Blanc: contents

Introduction 4

1 Planning and preparation
Planning to walk the TMB 5
Elevation and pace 6
Waymarking and variantes 7
Safety and weather 8
New to long distance walking? 8
Clockwise or anticlockwise? 9
How long will it take? 9
Travel planning 10
What is the best time of year? 11
Accommodation 11
Advance planning checklist 12
Packing checklist 13

2 Background information
2.1 History 14
2.2 Climbing Mont Blanc 15
2.3 Geology and glaciers 18
2.4 Habitats and wildlife 19

3 The routes in detail
3.1a Les Houches to Les Contamines via Bionnassay 22
3.1b Les Houches to Les Contamines via Col de Tricot 25
3.2a Les Contamines to Les Chapieux 27
3.2b Les Contamines to Refuge des Mottets via Col des Fours 30
3.3 Les Chapieux to Rifugio Elisabetta 32
3.4 Rifugio Elisabetta to Courmayeur 34
3.5 Courmayeur to Rifugio Bonatti 38
3.6 Rifugio Bonatti to La Fouly 40
3.7 La Fouly to Champex 42
3.8a Champex to Trient via Bovine 44
3.8b Champex to Trient via Fenêtre d'Arpette 46
3.9 Trient to Tré le Champ 48
3.10 Tré le Champ to La Flégère 51
3.11 La Flégère to Les Houches 56

4 Reference
Accommodation list 59
Books, maps, contact details and weather 61
Transport, acknowledgements and credits 62
Index 63

Introduction

Mont Blanc reflected in one of the Lacs des Cheserys

The Tour of Mont Blanc is one of the world's classic long walks. Otherwise known as the Tour du Mont-Blanc (or TMB), its heart is the sprawling glaciers and majestic rock spires of Mont Blanc itself. At 4810 m this is the highest peak in the Alps and the roof of Western Europe. More than a mere mountain, this massif hosts dozens of glaciers and hundreds of summits, and is surrounded by seven major valleys.

The TMB pursues an oval route around its epicentre, winding for some 170 km in and out of the valleys, linked by a succession of 10 or 11 high passes, gaining and losing some 10,000 m of altitude. Traditionally the TMB starts and finishes near Chamonix, the world capital of mountaineering. From there it makes its way through charming Alpine valleys, past lakes and glaciers and across the frontiers of France, Italy and Switzerland.

It's difficult to single out highlights of this route because it includes such variety. For sheer scenic grandeur, the days either side of the Italian town of Courmayeur, and the days above Chamonix around La Flégère, Lac Blanc and Le Brévent are perhaps unequalled in the Alps. But there is much more to the route than spectacular mountain views. Even during a spell of bad weather, there is always something to take pleasure in, whether it's the rushing of melt-water streams, the meadows of wildflowers, the tinkling of cowbells or the warm fellowship of a convivial meal in a mountain refuge.

1 Planning to walk the TMB

The TMB is undeniably a challenging walk, although there are various options for making the route easier and shorter. Cable-cars and chair-lifts can be used to skip difficult climbs and descents, and various sections can be bypassed. If relying on cable-cars to save time, be aware of seasonal opening months (generally June to September, but they vary) and last descent times (usually 5 or 5.30 pm) and check in advance with Chamonix Tourist Office: see page 61.

When planning your walk, make an honest assessment of your fitness and then study this guidebook carefully in order to plan your itinerary. It's better to be slightly conservative than over-ambitious. You'll want to have energy left over to enjoy the mountains, so allow a spare day or two in case of bad weather or the need for a rest day. So although it can be done in eleven daily stages, it is more realistic to allow up to two weeks if you are determined to complete the entire walk. If you are interested in also climbing Mt Blanc itself, see pages 15-17 and allow several extra days.

The TMB is serviced by dozens of well-situated mountain huts and *gîtes,* which not only offer a bed for the night but will also provide meals. This is a real bonus given the route's length and the amount of ascent and descent, because you need only carry a light daypack. You can tackle the walk in stages that suit your pace and fitness.

The greatest freedom comes from being totally self-sufficient and carrying everything (tent, food and cooking gear) on your back: to do this you need to be both experienced and fit. If you want to walk the TMB outside the normal season, you'll have no choice: see page 11.

Lac Blanc and the Chamonix Aiguilles

The TMB passes through several villages and small towns so there are plenty of opportunities to stock up on food. The longest section without full services is that between Les Contamines and Courmayeur, about three days. Even then you can still buy bread, cheese and other provisions from the small farm shops in Les Chapieux.

The TMB is a popular route for organised guided groups. Before booking to join such a group, clarify the exact itinerary. Many guided tours skip large chunks of the route, or detour around some of the best sections. Find out whether the package includes transfer of your baggage between some or all accommodations.

Elevation and pace

The TMB dips below the 1000 m contour only around Les Houches, and for the remainder of its journey stays well above 1500 m. The highest point is at 2665 m, and most of the passes are well above 2000 m. In all there is over 10,000 m (33,000 ft) of ascent and descent to negotiate, a daily average of almost 1000 m. Some stages are harder than others: on certain days you'll be climbing or descending up to 1600 m (5250 ft).

TMB altitude profile

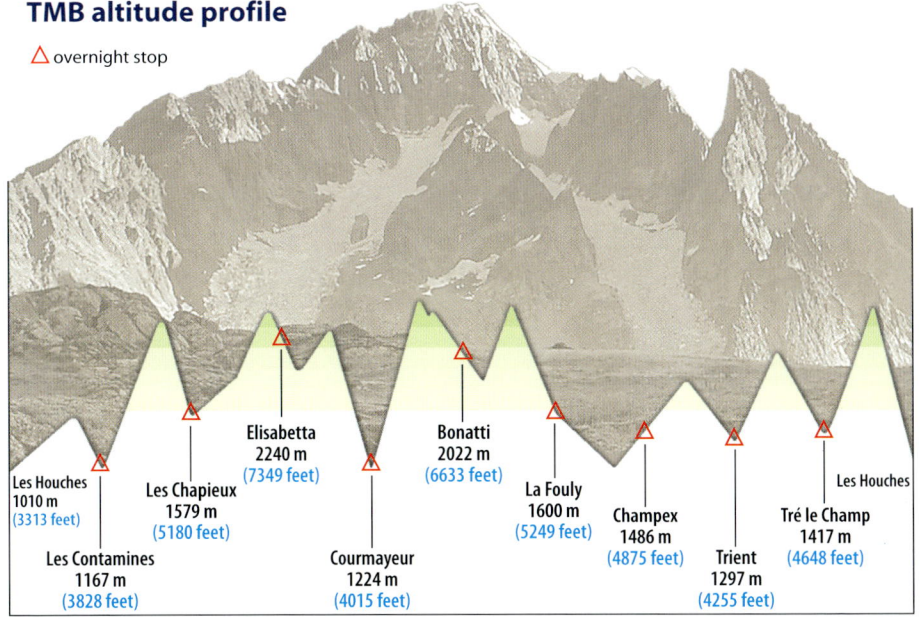

Generally speaking the paths and tracks used by the TMB are well constructed and allow for reasonably good progress, but in places more rugged terrain can reduce your average speed. Your speed will also vary considerably depending on whether you're on a long climb or a descent. Overall, expect to average 3-4 km/hr (2-2½ mph) unless you're particularly fit and impatient to press on. It's important not to go too fast in the early stages and to conserve energy until you have a good idea of your fitness relative to the difficulty of the walk. If you walk in a group, expect to travel at the pace of the slowest member or a bit less: groups tend to make more stops than individuals.

Waymarking and variantes

Signpost on Col Sapin, Italy

The TMB is generally well signposted, but the markers vary from country to country, sometimes even from valley to valley. Around the Chamonix Valley red and white paint on signs and splashed on rocks are used. In Italy, yellow and black diamonds with the letters TMB are painted on rocks and walls, whilst Switzerland uses the same diamonds but without the lettering. Elsewhere you'll simply be looking for the normal trail signs of that area. Bear in mind that the occasional marker may be missing or hard to see. In short, you need to watch carefully and check the map frequently to make sure you are on the right path.

The exact routeing of the TMB varies slightly from time to time, and you'll find inconsistencies between the marked route on the ground and the route shown in topographic maps. Discrepancies don't always matter much: in some places several equally scenic paths all go to the same destination.

Waymarker, Val Ferret, Switzerland

You'll also notice that there are official variations on the route known as variantes that are often harder than the original route. They are not necessarily more scenic, but often they're wilder. Some simply provide the option of visiting a different valley or spending the night in a different location. Three variantes are described as stages in their own right: see sections 3.1b, 3.2b and 3.8b, with other choices summarised in panels.

Safety and weather

Few sections of the TMB are dangerous or exposed, and taken with care they pose little problem to the vast majority of walkers. Early in the season, banks of snow on the higher ground can be dangerous, especially early in the morning. If the slope is steep and the surface icy, wait for the sun to soften the snow. In softer snow, be wary of boulders hidden beneath the surface. Some paths are quite rocky and if you are new to Alpine walking, be aware of the danger of a twisted ankle or worse.

The weather in the Alps can turn nasty at any time of the year: always check the forecast, see page 61. On the high passes, conditions can quickly become life-threatening, and even in the height of summer you may experience snowfall at altitudes above 2000 m. Thunderstorms are common in the afternoon, particularly in June and July, so it is always advisable to start early. That way you'll be walking in the cool of the morning when the views are clear. You're also more likely to be safe in a refuge, or at least past the day's high point, should a thunderstorm strike.

Avoid dehydration by carrying plenty of drinkable water. Although many water sources on the trail are fine, the safe policy is always to purify before you drink. Carrying iodine (drops or tablets) is usually more convenient than boiling.

If you're walking alone you should consider carrying a mobile phone in case of emergencies, bearing in mind that coverage is patchy. Consider also carrying a space-blanket type of survival bag: it weighs almost nothing and could save a life in the event of an emergency.

Learn the internationally recognised distress signal: six short signals (whistle, shout or flash of light) at 10-second intervals, followed by a one minute pause. The response is three signals at 20-second intervals with a minute between each.

New to long-distance walking?

If you haven't done much walking before, the TMB would be an ambitious first choice. If you are fit and keen it is feasible, but don't underestimate the importance of looking after your feet. An unattended blister on day one could cause you severe lameness or even to have to curtail your holiday. At minimum, your preparation should include several strenuous walks of at least 6-8 hours, preferably with a TMB level of ascent and descent, ideally on consecutive days. Wear the same footwear as you intend to use for the TMB. Socks and footbeds are almost as important as the boots themselves.

Assess your fitness objectively, long before you leave for the Alps. Allow yourself time to experiment with alternatives. If in doubt, consider fallback positions, for example using cable-cars and chair-lifts. Think about how, if you were really suffering, you might curtail the hike. For advice on hiking gear and on blister prevention and treatment obtain our *Notes for novices*: see page 62.

Clockwise or anti-clockwise?

The TMB is traditionally walked in an anti-clockwise direction for several reasons, notably orientating the best views to face the walker, a better positioning of long, difficult climbs and slightly more obvious waymarking. Therefore we describe the walk in this direction. You could walk the TMB clockwise instead.

The main advantage is that you would avoid walking with the crowds during the day. However you'll have to pass them at some point and the refuges will be equally busy, although you'll see new faces each evening. If you start in Les Houches and walk clockwise you'll face over 1500 m of ascent on the first day.

How long will it take?

In this guide the TMB is broken down into eleven stages, each a manageable day's walk for someone of reasonable fitness carrying a modest pack. You could cut a day or two off the total using cable-cars and taking the bus between La Fouly and Champex. If you adopt this approach you could also skip the days above Chamonix and make a somewhat truncated Tour of Mont Blanc, starting from Les Houches in the Chamonix Valley and finishing in Tré le Champ, also in the Chamonix Valley, within about seven or eight days.

Although this might be the only option for those with limited time, it is certainly not the best approach. Ideally you should allow spare days in case of bad weather, to have a rest or to enjoy a refuge or location that you particularly like. Some walkers also break the stages given here into even smaller chunks and take more than two weeks to complete the circuit. It is because exploration is more important than completion that this book is called *'Explore* the Tour of Mont Blanc'.

Table 1: The TMB (anti-clockwise) in 11 daily stages

Section	From	To	Distance (km)	(miles)	Ascent (m)	(feet)	Descent (m)	(feet)
3·1a*	Les Houches	Les Contamines	16	10	890	2920	730	2395
3·2a*	Les Contamines	Les Chapieux	19	12	1300	4265	930	3050
3·3	Les Chapieux	Rifugio Elisabetta	14	9	1000	3280	400	1310
3·4	Rifugio Elisabetta	Courmayeur	15	9	490	1610	1400	4595
3·5	Courmayeur	Rifugio Bonatti	13	8	1600	5250	680	2230
3·6	Rifugio Bonatti	La Fouly	18	11	880	2885	1420	4660
3·7	La Fouly	Champex	15	9	430	1410	575	1885
3·8a*	Champex	Trient	18	11	780	2560	730	2395
3·9	Trient	Tré le Champ	14	9	960	3150	830	2725
3·10	Tré le Champ	La Flégère	7	4½	790	2590	330	1085
3·11	La Flégère	Les Houches	18	11	770	2525	1615	5300

* data is for the main TMB; variantes are described in book sections 3·1b, 3·2b and 3·8b.

Flags and flowers lining the river, Chamonix

Travel planning

The traditional start at Les Houches is very accessible, just a few kilometres from the holiday centre of Chamonix on the north-western edge of the Alps. The closest international airport is Geneva just 100 km away. Several major airlines have scheduled services to Geneva, including some low-cost operators. See page 62 for contact details for transport by air, rail and bus.

Both train and bus services link Geneva International Airport to Chamonix. The local SAT Autocar bus service takes about 2½ hours, and stops in Les Houches. There are also quicker but more expensive minibus services that will drop you to the door of your accommodation. Taking the train is the cheapest option, but you'll need to change twice and the journey takes about three hours.

Another option is to fly to Turin, which is only three hours by bus from Courmayeur. You could then either take another bus through the Mont Blanc tunnel to Chamonix (45 minutes), or simply start and finish in Courmayeur.

What is the best time of year?

From the end of October until the start of June the high passes of the TMB are blocked by snow. The walking season begins in June, with the refuges opening about the middle of the month, although there will still be banks of snow on the high parts of the TMB. The amount of snow persisting into the early summer varies from year to year: check conditions with the tourist office or Office de Haute Montagne in Chamonix, see page 61. Occasionally, an unusually large snowpack and a poor thaw can cause problems for early-season walkers.

Late June and early July bring warmer weather and relatively uncrowded conditions. From mid-July until the end of August it's the high season, with many people on the trail and the possibility of accommodation being full. In some years the weather can be uncomfortably hot for walking.

September is a great month: the crowds have died down, but the refuges are still open and you can often enjoy long spells of fine weather. October can be wonderful too, particularly the first couple of weeks, with the trees in their autumn finery and the first snows of winter plastered on the high peaks. However you'll need to plan your accommodation carefully at this time of year, or else carry a tent, as most refuges and gîtes are closed.

Accommodation

Apart from places such as Courmayeur, Champex and Les Contamines where you could stay in four-star luxury, accommodation is mainly in mountain huts (*refuge* in French, *rifugio* in Italian) and gîtes. Although their facilities are modest, they always supply blankets, allowing you to save weight and space by carrying only a liner bag, rather than a full sleeping bag.

Sleeping arrangements are communal, in dormitories with individual beds or in bunkrooms with mattresses side-by-side on wooden platforms. Don't expect privacy or a peaceful night's sleep. Although everyone goes to bed early, refuges can be noisy places, complete with heavy snorers. Earplugs are recommended

Refuges on the TMB serve hot food and drinks, and it's normal to pay for half-board (demi-pension). In 2004, the cost of this was about €40 per person, for which you get your bed, a filling evening meal, and breakfast. Some refuges also offer a packed lunch for the following day's walk. Few have kitchen facilities for walkers, so don't expect to cook your own meals.

Guest house and mountain view, La Fouly, Val Ferret

Most of the refuges on the TMB are privately owned, but some are operated by the French or Italian alpine associations. They offer a discount of up to 50% on the cost of accommodation to members of the British Mountaineering Council – and to other national mountaineering bodies with reciprocal agreements .

The refuges are generally open from mid-June until the end of September. During July and August it is essential to book in advance. Walkers turning up unannounced will receive little sympathy if the refuge is full. The French authorities produce a TMB accommodation listing every year, detailing prices, contact numbers and opening times. Request this from tourist information offices on the French side of Mont Blanc or from the Office de Haute Montagne in Chamonix: see page 61 for contact details.

TMB refuges run by the alpine associations have winter rooms that can be used by out-of-season walkers. In the case of the Refuge du Col de la Croix du Bonhomme (stage 2) the winter room has a gas cooker, a wood burner and beds set up in the spacious dining area, which also has solar-powered under-floor heating. The Rifugio Elisabetta has a simple annex room with bunks and electric lighting, while the Rifugio Elena has a similar room but without the lights.

Gîtes are essentially hostels, with accommodation in dormitories and a communal kitchen. Unlike refuges, meals are not provided, but since most gîtes are located in villages and towns you can always buy your own food.

Camping is another option, but before committing yourself you should consider all the implications of carrying a heavy pack. Most of the villages and towns on the route have campsites with facilities. Many of the refuges also permit campers to pitch nearby for free, and you'll be welcome to use their facilities and have meals provided that the refuge is not too busy. Special provisions apply within the boundaries of the Reserve Naturelle des Aiguilles Rouges (stage 10), where between the hours of dusk and dawn only bivouacs are officially permitted. In theory this means tents with poles and pegs are not allowed, but in practice discreet overnight camping well away from the Lac Blanc refuge seems to be tolerated.

Advance planning checklist

- Consult medical advisor about your proposed trip.
- Plan and execute training programme for fitness and stamina.
- Take out suitable insurance as soon as you book; must cover search and rescue.
- Learn some French phrases for booking accommodation, ordering food and asking directions.
- Book accommodation as soon as you know your itinerary.
- Weigh all your kit and decide whether to buy or borrow any lightweight upgrades.

Packing checklist

The checklist is divided into essential and desirable items, and is intended to be used as a starting point. It assumes that you plan to use refuges and gîtes where blankets or duvets are provided. If your boots and waterproofs haven't been worn recently, test them before you go, while there is still time to re-proof or replace them if need be.

If you are camping, you will have to carry more weight (at least an extra 5 kg) to include a tent, warm sleeping bag and mat, food, cooking utensils, a portable stove and fuel – and a much larger, heavier rucksack to carry it all.

Essential

- rucksack with plenty of room for everything (minimum 35 litres)
- waterproof rucksack cover or liner(s), e.g. heavy duty bin bag
- comfortable walking boots (preferably waterproof)
- suitable clothing in layers, with specialist walking socks
- waterproof jacket and over-trousers
- hats (for warmth and sun protection) and gloves
- dry clothes and footwear to change into
- sun block with high Sun Protection Factor
- good quality sunglasses
- sleeping bag liner (silk, cotton or thermal)
- torch (preferably head-torch) and spare batteries
- water carrier (bottle or bladder) and iodine purification tablets or drops
- first aid kit including blister treatment
- toilet tissue (biodegradable)
- personal toiletries and towel
- guidebook and map
- more than enough food to last between supply points
- enough cash in euros; banks in Chamonix, Les Houches, Les Contamines, Courmayeur, Champex, and Argentière have cash machines; credit cards are widely accepted, but don't rely on them.

Desirable

- trekking poles: two are recommended
- light and rugged camera, plus spare batteries, film and memory cards (if digital)
- mobile phone (useless unless you can conserve its battery level)
- whistle and survival bag (essential if hiking alone)
- pouch or secure pockets for keeping small items handy
- binoculars – useful for watching wildlife
- notebook and pen
- earplugs
- down jacket (in autumn).

2·1 History

Some of the paths you'll follow on the TMB have been used for millennia. Five centuries before the Romans arrived, Celtic tribes were already using the Col du Bonhomme as a trade route for salt. The Romans continued to use this pass and you'll walk a section of old Roman road, and use a Roman-built bridge, on the lower slopes of the Col du Bonhomme. Many tracks began as shepherds' routes, used for trade between the valleys, with meat, cheese and other dairy produce being the main export.

Exploration of the peaks didn't really begin until the late 18th century when a handful of men began to debunk the commonly held beliefs that the high mountains were the preserve of evil spirits and demons.

Three names are famous in the ascent of Mont Blanc and the birth of mountaineering. Horace-Benedict de Saussure, a professor of physics from Geneva, put up a reward for the first ascent of Mont Blanc. In 1786 two Chamonix men – Dr Michel Paccard and a chamois and crystal hunter named Jacques Balmat – were the first to reach the summit.

Saussure himself claimed the third ascent, and also made the first circuit of the mountain. From a starting point in Chamonix he reached Courmayeur by a route generally similar to the modern TMB. Thereafter he followed a more circuitous route into Switzerland, crossing the Grand St Bernard Pass.

The first ascent of Mont Blanc by a woman was made in 1808 by Marie Paradis. Throughout the 19th century the more technically difficult peaks around Mont Blanc fell to Victorian enthusiasts such as Edward Whymper and local guides such as Michel Croz. This era was later referred to as the 'Golden Age' of mountaineering.

Aspects of Mont Blanc provided new challenges until the modern era. In 1961 the Frêney Pillar on the Italian side of Mont Blanc claimed the lives of four men from a party led by the famous Walter Bonatti (see page 40). Only weeks later the first ascent of this route fell to Chris Bonington and Royal Robbins.

In the 1960s and 1970s Alpine walking began to increase in popularity. The refuges dating from the Golden Age were expanded to cope with the numbers, and new accommodations sprang up in the valleys. Paths such as the Grand Balcon Sud, built for walkers in the late 19th and early 20th centuries, were extended and improved. The TMB as we know it today had arrived.

Above: statue of Balmat and Saussure, Chamonix

2·2 Climbing Mont Blanc

Climber on Mont Blanc

The Refuge du Goûter

By modern standards Mont Blanc is a relatively easy ascent. Climbers with mountaineering skills (including crevasse rescue and experience with ropes, crampons and ice-axe) could consider an unguided ascent in good weather. Fit walkers who lack these technical skills should hire a guide or book with an organised group: for contact details see page 61.

Proper equipment is essential for safety. Take an ice-axe, crampons, climbing helmet, rope and harness. Plastic mountaineering boots are desirable, even in summer. If you rely on your hiking boots, make sure that they are waterproof, warm and rigid enough to take crampons securely. Poles are useful for crossing the Dome du Goûter.

You'll also need to prepare your body by acclimatising beforehand. Aim to spend at least two nights above 3500 m before your summit attempt and be alert for any symptoms of altitude sickness. This is mainly a high-altitude snow plod, but on a high and complex mountain the weather can change rapidly and any mistakes in navigation could lead to disaster. Attempt the ascent only in settled weather.

Camping on the Aiguille du Goûter

The easiest and most popular route to the summit is via the Refuge du Goûter. The Bellevue cable-car takes you from Les Houches up to the Col de Voza, then the Tramway du Mont Blanc funicular continues to the Nid d'Aigle at 2372 m. From there, walk to the bottom of a couloir and cross it with great care, wearing a helmet. Many climbers have been killed or injured by falling rocks on this short section. Then there's a long scramble up a rock rib to reach the Refuge du Goûter at 3817 m. (If you prefer to carry a tent and suitably warm overnight gear, you can camp just above the Refuge.)

After overnighting here, aim for a 2 am start to make the most of the frost-hardened snow. A long slog up and over the Dome du Goûter leads to the Vallot emergency hut where the ridge narrows. From here to the summit is the most difficult section of the route, along an exposed knife-edge of snow that often forces climbers to turn back because of high winds and altitude. However, for those who are lucky with conditions, the views from the summit are immense, with a panorama of hundreds of peaks.

Descending Mont Blanc

2·3 Geology and glaciers

The geology of the Mont Blanc area illustrates the dynamic and complex geology of the Alps as a whole. Around the fringes of the mountain, and in many of the areas where TMB walkers will find themselves, the rock types are principally limestone and schist. These are the result of a rock-building process that took place almost 700 million years ago. The Mont Blanc massif itself and the Aiguilles Rouges are composed of granite, which pushed up through the older rocks some 300 million years ago. The actual tectonic process of mountain building only began about 100 million years ago when the African and European plates collided and forced these rocks upwards.

The period of mountain building in the Alps is thought to have ended some six million years ago and since then the effect of successive periods of glaciation have created the landforms we see today. However Alpine glaciers are currently in rapid retreat and have been for over a century. In 1645 the Bishop of Geneva performed an exorcism at the snout of the Mer de Glace, after it had bulldozed through chalets and farmland in the Chamonix Valley. This glacier, the second longest in the Alps, has since then retreated several kilometres.

Glaciers, on the north face of Mont Blanc

2·4 Habitats and wildlife

The TMB passes through three major alpine habitats. The montane zone is the lowest of the three, characterised by mixed broadleaf woodland and animal pastures. At about 1200 m this gives way to the sub-alpine zone, dominated by pine forests: principally Norway spruce, European silver fir and the deciduous European larch whose needles turn a rich golden colour in autumn. In rocky zones or where avalanches are common, you'll find swathes of birch.

Moss campion

The alpine zone begins above the tree line, although its altitude varies significantly from about 1200-2000 m. Generally though the thinning out begins at about 1700 m, and few trees survive much above 2000 m. Those that do are more like shrubs, including several species of dwarf willow. Indeed shrubs are the main constituents of the alpine heath that dominates the open slopes above the tree line. Among these shrubs you'll find bilberry, which produce a profusion of tangy berries and wonderful red leaves in autumn.

Alpine gentian

Male ibex browsing in autumn foliage

Higher still in the alpine zone the shrubs give way to hardy grasses, mosses and lichen, and the occasional wildflower that can survive at extreme elevations such as the Alpine gentian. Unfortunately for lovers of wildflowers, many of the open slopes are heavily grazed and don't produce the profusion of flowers that can be seen in other parts of the Alps. Having said that you can expect to come across some good pockets of colour.

Perhaps the largest and most charismatic mammal you're likely to see is the ibex. Males have very distinctive scimitar-like horns. Once hunted to the brink of extinction, they are now making a comeback in parts of the Alps. Your best chance of seeing ibex is on the stage from Tré le Champ to La Flégère where the herds resident in the Reserve Naturelle des Aiguilles Rouges are tolerant of humans and can often be seen very close to the path. Much more skittish are chamois, the ibex's delicate small cousin, with distinctive forward-curving horns. They too are most likely to be sighted around the Reserve Naturelle des Aiguilles Rouges, particularly early in the morning.

Marmots are the cutest of Alpine mammals. They are found above the tree line where they make burrows in banks or in boulder fields. These large rodents signal alarm to other marmots with a shill whistle, and this is often your first clue to their whereabouts. Following a warning call you can often see them scurrying towards the safety of their burrow.

Commonly seen birds, particularly in the proximity of refuges and picnic sites, include ravens and their rarer, yellow-billed cousins, choughs. In the mixed habitat offered by valleys like the Val Ferret you also stand a good chance of seeing birds of prey including buzzards and eagles. The shy ptarmigan can often be heard rather than seen. An insistent clucking is often the only hint of its presence amongst the thickets of bilberry just above the tree line.

Marmot

3·1a Les Houches to Les Contamines

via Bionnassay

Map	panel 1	**Distance**	16 km (10 miles)	**Time**	5-6 hours

Terrain   mostly tracks and small roads

Grade   steep ascent to Col de Voza; steady descent to Tresse interrupted by two short climbs; gentle ascent to Les Contamines

Total ascent   890 m (2920 ft)   **Total descent**   730 m (2395 ft)

Food and drink   Les Houches, Col de Voza, Bionnassay, Les Contamines

Summary   this stage has more tracks and tarmac than others, but views are good and you pass through pretty Alpine hamlets; use the cable-car if you want to skip the first ascent

From the centre of Les Houches, walk west along the main road for 1 km until you reach the Téléphérique Bellevue. To ascend by cable-car, see the panel on page 23. Otherwise continue past the cable-car station for 30 m and turn left under a small bridge, following a sign for the Col de Voza.

Climb steeply through woodland on a vehicle track and then turn right on to a surfaced road. After only 100 m turn left on to another road and follow this as it switchbacks uphill past several chalets. You gain good views of the Chamonix Valley as you climb.

The ascent to Col de Voza

Col de Voza

At the end of this road turn right on to a track and 100 m later turn right again, this time passing under a chairlift. Don't be put off by the sign in French prohibiting non-forestry vehicles: it doesn't apply to walkers.

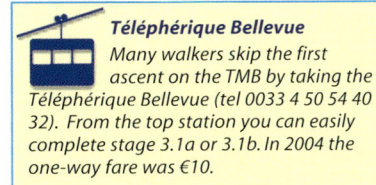

Téléphérique Bellevue
Many walkers skip the first ascent on the TMB by taking the Téléphérique Bellevue (tel 0033 4 50 54 40 32). From the top station you can easily complete stage 3.1a or 3.1b. In 2004 the one-way fare was €10.

Now follow a steep track through woodland and pastures to a small restaurant. From here the track continues to climb steeply, between woodland on the right and winter ski pistes on the left, reaching the Col de Voza (1653 m) after another 45 minutes. The broad col is dominated by ski buildings, but provides excellent views of Mont Blanc and west across the foothills of the Alps.

Turn left by a large hotel and after 100 m turn right and cross the tracks of the Tramway du Mont Blanc. Turn left here for the alternative route via the Col de Tricot (section 3.1b). Otherwise, bear right on to a vehicle track signed for Bionnassay.

The track descends steeply, reaching the Refuge du Fioux after 20 minutes. Descend past this refuge, enjoying good views of the Glacier de Bionnassy, and then turn left on to a surfaced road. Continue along this road for 500 m to the hamlet of Bionnassay, with the Auberge de Bionnassay just off to the right.

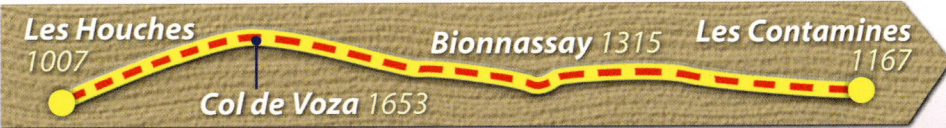

The TMB turns left on to a small footpath, descending past a tiny church and through woodland to reach a footbridge across the Torrent de Bionnassay. On the opposite bank, climb steeply through switchbacks to a forest track where you turn right.

Follow the track round a forested spur and down into the hamlet of Le Champel. Surfaced roads lead down through the hamlet, after which you turn left at a hairpin bend on to the Chemins des Chevreuils. This small track descends steeply through woodland to reach another surfaced road.

Turn left on to the road and follow it uphill past the chalets of La Villette. Use a small path on the left to shortcut a hairpin in the road. The road then descends to a small bridge across the Torrent de Miage. Just upstream from the bridge are the oppressive rock walls of the Gorges du Gruvaz.

On the other side of the bridge turn immediately right on to a small path. This descends through woodland into the hamlet of Tresse, where it meets a minor road. Follow the road to a junction with the main D902 Les Contamines-St Gervais road.

Cross the main road and continue along a minor road, passing over a bridge and climbing round a hairpin bend. The road becomes a track, and then a path that descends slightly to join a surfaced road in the hamlet of Les Hoches.

Follow this road for a few hundred metres and cross a bridge. The TMB now goes off to the right, following a pleasant, wooded, riverside path. After less than 2 km, a final steep climb brings you to the square in the centre of Les Contamines, where you'll find a bakery, a supermarket, and the tourist information office.

Alpine church, Les Contamines

3·1b Les Houches to Les Contamines via Col de Tricot

Map	panel 1
Distance	18 km (11 miles)
Time	7–8½ hours
Terrain	tracks and roads to Col de Voza then good paths to the Refuge du Truc; mostly track for the descent into Les Contamines
Grade	steep ascent to Col de Voza, then steadier to Col de Tricot; steep descent to Miage; a further climb before the final long descent
Total ascent	1480 m (4855 ft)
Total descent	1340 m (4400 ft)
Food and drink	Les Houches, Col de Voza, Refuge de Miage, Refuge du Truc, Les Contamines
Summary	a very tough stage, but can be made much easier by using the cable-car for the initial ascent; fine views of the west side of Mont Blanc, with an exciting swing bridge to cross

Follow directions in stage 3·1a as far as the tramway on the Col de Voza. There you turn left along a wide path, following signs for the Col de Tricot (and TMB).

After 1 km bear right along a forest path, contouring high above the Bionnassay Valley. After a further 1 km, the path descends on to the terminal moraine of the Glacier de Bionnassay. Follow cairns through the boulders to a swing bridge.

Swing bridge over the Torrent Bionnassay

Once across the bridge the route climbs steeply up scrubby slopes and then eases back a bit as it continues up to the Col de Tricot (2120 m). Views are constrained here by the steep ridges on either side, but there is an airy perspective on the hamlet of Miage over 400 vertical metres below.

Descend steeply on a series of switchbacks to reach the hamlet and the prominent Refuge de Miage. Cross the Torrent de Miage on a footbridge and begin a steep climb that brings you to a broad shoulder in slightly over half an hour.

The path now descends gently past the cosy-looking Refuge du Truc, where it joins a vehicle track. The track drops steeply into pine forest. Look out carefully for a path on the left (a shortcut). Follow this path for a short distance to a junction where you turn right, descending back on to the vehicle track.

Follow the vehicle track as it descends steeply through the chalets above Les Contamines. To reach the centre of the village continue straight ahead on the Chemin du P'tou.

Col de Tricot

3·2a Les Contamines to Les Chapieux

> **Map** panel 2   **Distance** 19 km (12 miles)   **Time** 6½-7½ hours
>
> **Terrain** wide tracks as far as the Refuge de la Balme and paths, sometimes rocky, for the remainder, with possible patches of snow beneath the Col du Bonhomme
>
> **Grade** ascent is long but generally steady, with steep sections beneath the Col du Bonhomme; steep paths down to Les Chapieux
>
> **Total ascent** 1300 m (4265 ft)   **Total descent** 930 m (3050 ft)
>
> **Food and drink** Les Contamines, Refuge de Nant Borant, Refuge de la Balme, Refuge du Col de la Croix du Bonhomme, Les Chapieux
>
> **Summary** a tough stage that can be split by staying overnight in the Refuge du Col de la Croix du Bonhomme; wonderful scenery with fine views of the Beaufortain Alps

From the centre of Les Contamines the TMB follows the main road south for several hundred metres before turning right on to a riverside path. The path leads gently up the valley, passing through a large public recreation area. About 30 minutes from Les Contamines you pass a sign for Camping le Pontet, 200 m off to the right.

A wide path continues up the Val Montjoie to the beautiful church of Notre Dame de la Gorge. The TMB now climbs steeply along a track, originally constructed by the Romans, which is roughly paved with large slabs and flagstones.

Church of Notre Dame de la Gorge, Les Contamines

Refuge du Col de la Croix du Bonhomme

Cross a stone bridge (also Roman) above the thundering Cascade de Combe Noire and continue to climb past a small chalet restaurant and the Refuge de Nant Borrant. This steep ascent brings you to a much wider and relatively flat upper valley. There is a free camping area just off the track on the left.

Follow the main track as it climbs gently up the valley to the Refuge de la Balme. There are toilets, a water fountain, and another free camping area just before you reach the refuge.

Beyond the refuge the TMB turns left on to a rocky trail, and climbs the steep valley headwall. The trail splits into many strands, all of which join up again as you climb into another broad valley.

A path to the left is signed for a diversion to the Lacs Jovet – a pair of alpine lakes tucked away in a basin a few hundred metres higher up. Allow an extra 1½ hours for this round trip.

Meanwhile the TMB continues along the floor of the valley and then swings right, climbing steeply once again. After 20 minutes the gradient eases, climbing steadily through a basin beneath the Col du Bonhomme. For much of the summer there may be a bank of snow to cross here.

Beyond this the trail climbs steeply to reach the Col du Bonhomme (2329 m). There is a small wooden shelter here and wide views to the south-west across the Beaufortain Alps.

The TMB now swings to the south and climbs across steep slopes and around a broad spur. Climb up through a small gully and then turn to the right to reach the cairn marking the Col de la Croix du Bonhomme (2479 m). The refuge of this name is visible just 200 m away, slightly beneath the col. The *variante* via the Col des Fours bears off to the north from here: see stage 3.2b, page 30.

The main TMB descends past the refuge on a braided path, dropping steeply through small gullies and across streams. After 20-30 minutes, an unmarked path goes off to the left. Ignore this and keep to the right, continuing steeply down to a shepherd's hut.

Beyond the hut the trail drops away again across open pastures, heading towards a group of buildings and a vehicle track. Turn left on to this track, and left again on to a path.

A combination of signed paths and tracks leads down into Les Chapieux, where you emerge opposite the Refuge de la Nova.

Sheep grazing above Les Chapieux

3·2b Les Contamines to Refuge des Mottets

Map panel 2	**Distance** 7 km (4½ miles) **Time** 7-8 hours

Terrain path across the Col des Fours that may be covered in snow well into the summer, narrow paths down into La Ville des Glaciers and then a track

Grade gentle ascent to the Col des Fours; then a long, steady descent to La Ville des Glaciers, and a final gentle climb

Total ascent 1550 m (5085 ft) **Total descent** 880 m (2885 ft)

Food and drink Refuge du Col de la Croix du Bonhomme, Refuge des Mottets

Summary this variante is wilder and more scenic than going via Les Chapieux, and not too difficult if you stay in the Refuge du Col de la Croix du Bonhomme

From the Col de la Croix du Bonhomme follow a signed path that climbs steadily to the north-east. After climbing for 20-30 minutes, you reach the Col des Fours (2665 m) where there is almost always a bank of unmelted snow.

A path breaks off to the left from the col, leading up along the obvious ridge to the cairn on the Tête Nord des Fours (2756 m). This side-trip takes about 30 minutes return and gives more panoramic views than those from the col.

Walkers below Les Aiguilles des Glaciers

Refuges des Mottets

On the east side of the Col des Fours the path descends steeply into a small, grassy basin. It then drops steeply again past slabs of rock to a poorly marked junction beside a small stream.

Turn right here and descend to a junction with a vehicle track. Turn right on to the track for just a few metres before turning right again on to a path.

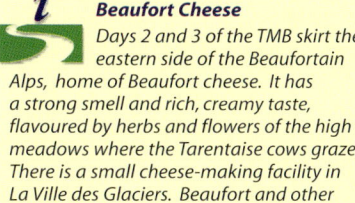

Beaufort Cheese
Days 2 and 3 of the TMB skirt the eastern side of the Beaufortain Alps, home of Beaufort cheese. It has a strong smell and rich, creamy taste, flavoured by herbs and flowers of the high meadows where the Tarentaise cows graze. There is a small cheese-making facility in La Ville des Glaciers. Beaufort and other cheeses are used in Savoyard specialities such as fondue.

Continue on this path as it cuts across the switchbacks of the track, descending directly to La Ville des Glaciers where you rejoin the main TMB. The nearest accommodation is at the Refuge des Mottets, 20-30 minutes further along: see above.

3·3   Les Chapieux to Rifugio Elisabetta

Map	panel 2	**Distance** 14 km (9 miles)	**Time** 4-5 hours

Terrain tarmac road for the first 4½ km, then a short distance on track; the rest of the stage is on a good path

Grade steady ascent as far as the Refuge des Mottets; long steep climb to the Col de la Seigne; easy descent to the Rifugio Elisabetta

Total ascent 1000 m (3280 ft) **Total descent** 400 m (1310 ft)

Food and drink Les Chapieux, Refuge des Mottets, Rifugio Elisabetta

Summary very scenic stage, with the tough climb to the Col de la Seigne rewarded by superb views; as you enter Italy, you gain fresh views of the south side of Mont Blanc

Follow the surfaced road north-east from Les Chapieux. The road climbs steadily through a narrow, steep-sided valley, reaching the summer cheese-making hamlet of La Ville des Glaciers in 1-1½ hours.

Although this road section can be monotonous, on clear days you glimpse Mont Blanc peeping over the Col de la Seigne.

Turn right on to a track and go across a bridge. After climbing steadily for 20-30 minutes along the track, you reach the Refuge des Mottets, once an old creamery: see page 31.

Descending from Col de la Seigne

From here the path climbs steeply in a series of switchbacks before veering north-east and climbing round a series of broad spurs, also crossing a couple of streams. Finally the path emerges on to relatively gentle slopes, but the climb to the top of the Col de la Seigne (2516 m) can still seem interminable.

This broad pass marks the border between France and Italy. There are spectacular views to the north-east across the many ridges and pinnacles on the south side of Mont Blanc.

The descent into the Vallon de la Lex Blanche begins gently, but soon becomes steeper as you drop down to cross a stream. The path joins a vehicle track, leading down to the pan-flat lower half of the valley. Within a few minutes you sight the Rifugio Elisabetta up on the left, reached by a short climb up a track.

Dawn over the Rifugio Elizabetta, Val Veny, autumn

Les Chapieux *1554*
Col de la Seigne *2516*
Rifugio Elisabetta *2240*

3·4 Rifugio Elisabetta to Courmayeur

Map	panels 2 and 3 **Distance** 15 km (9 miles) **Time** 4½-5½ hours
Terrain	tracks for the first few kilometres, then good paths all the way to Courmayeur
Grade	initial descent followed by flat walking along the Val Veny; then a steep climb and a gradual descent to the Col Chécrouit; finally a long descent
Total ascent	490 m (1610 ft) **Total descent** 1400 m (4595 ft)
Food and drink	Rifugio Elisabetta, Rifugio Maison Vieille, Dolonne, Courmayeur
Summary	tremendous views on this stage looking across to the glaciers and rock faces on the south face of Mont Blanc; it makes sense to use the cable-car for the descent to Courmayeur

Beneath the Aiguille Noire de Peutery

The south face of Mont Blanc reflected in a pool above Val Veny

Descend from the Rifugio Elisabetta to pick up the TMB, and follow the vehicle track as it descends by a series of long switchbacks. Several rough paths cut the corners. At the bottom of this descent the track runs flat, and mostly dead straight, for the next 2 km along the south side of the valley floor.

Pass Lac de Combal and look for a path going off to the right, signed for the Col Chécrouit. The path climbs steeply through trees and shrubs, and then emerges on to open hillside at a ruined shepherd's hut.

Continue climbing along the lip of a steep ravine and then beside the left (east) bank of a stream. Keep left at a junction and keep climbing to another shepherd's hut.

The path now veers north-east and climbs round a spur, to a cairn marking the high point of this stage (2365 m). Enjoy superb views across the Val Veny to the south face of Mont Blanc: the Glacier du Brouillard and Glacier de Frêney fall dramatically from the buttresses of the Frêney Pillar in a twisted series of crevasses and seracs.

The path descends to a stream, climbs across a slight rise and then drops down past a small lake. On calm days the still waters give wonderful reflections of Mont Blanc.

Autumn woodland beneath the Aiguille des Glaciers, Val Veny

Follow the path as it descends steadily into scattered larch forest. The pistes and chairlifts of the Courmayeur ski area now become obvious. The path steepens, taking you down through thicker forest and out on to the Col Chécrouit (1956 m). Walk past the Rifugio Maison Vieille and follow a dusty track towards the obvious chairlift: see panel for this descent method.

Follow the track round to the right and just before it starts to climb again, turn left on to a small path. This path drops steeply down past chalets and along the edge of some trees. At the top of a fenced-off ravine, walk left to the top station of the Courmayeur cable-car on the Plan Chécrouit (1697 m).

> **Courmayeur cable-car**
> *There is little to recommend the long descent into Courmayeur, except for the narrow alleys of Dolonne, which can just as easily be visited from Courmayeur. The quick way down is by cable-car (with an upper section by chairlift): tel 0039 165 8 99 25). In 2004 the one-way fare was €8.*

To descend on foot, walk left past the station and take a narrow path that descends a steep slope on a series of switchbacks. The path joins a dusty vehicle track for a short distance. Watch carefully for where it turns off to the left, on the apex of a hairpin bend (signed, but easily missed). This path now descends steeply through forest to the edge of Dolonne, where you join a narrow surfaced road.

Follow yellow and black marks painted on walls as the TMB winds through the village's narrow alleys. Turn left on to a main road, go across a bridge and follow the road through an underpass to emerge in Courmayeur's Piazzale Monte Bianco. The tourist information office is in the square and there is a small supermarket nearby.

Courmayeur via the Val Veny

From the Col Chécrouit you can turn left and descend into the Val Veny to make a longer approach to Courmayeur. This takes you past the Rifugio Monte Bianco, with the option of a detour to the campsites in the Val Veny. The last 4-5 km into Courmayeur are along a road passing the terminal moraine of the Glacier de la Benva.

Glacier de la Lee Blanche, Mont Blanc massif

3·5 Courmayeur to Rifugio Bonatti

Map	panel 3 **Distance** 13 km (8 miles) **Time** 6-7 hours
Terrain	road and track at first, then paths all the way, rough at times – particularly on the descent to the Col Sapin
Grade	steady ascent for the first 2 km, then a long, steep climb to the Tête de la Tronche; short, steep descent to the Col Sapin, then an easier section and a final gentle descent
Total ascent	1600 m (5250 ft) **Total descent** 680 m (2230 ft)
Food and drink	Courmayeur, Rifugio Bertone, Rifugio Bonatti
Summary	one of the hardest stages of all, rewarded by incredible views and some airy ridge walking; the stage can be split over two days with a night at the Rifugio Bertone, or even avoided

Pick up the TMB as you leave Courmayeur, following signs up Strada del Villair, to the left of the church. A surfaced road climbs steadily past the chalets of Villair, a small hamlet.

From Villair the road becomes a vehicle track, and after 15-20 minutes it swings left across a bridge. Follow TMB signs to the right, climbing steeply through stands of birch to rejoin the vehicle track after its long hairpin bend.

After about 200 m the main TMB turns left off the track, on to a path that zigzags steeply all the way up to the tree line, whereas the Val Sapin *variante* continues straight ahead, see panel.

Just above the trees lies the charming Rifugio Bertone, some 2-2½ hours from Courmayeur. There are good places for camping just above it.

Across the Val Veny to Mont Blanc's south face

Monte Saxe with the Grandes Jorasses

The path now bears north-east and climbs steeply up the shoulder of Mont de la Saxe. The gradient eases and after some avalanche fences the path drops slightly. It then climbs underneath the Tête Bernarda and on to a wonderful section of ridge leading to the Tête de la Tronche (2584 m). In clear weather, this has tremendous all-round views, notably across the Val Ferret to the mighty south face of the Grandes Jorasses.

A short but very steep descent leads down to the Col Sapin (2435 m). Care is needed in places as the trail is quite exposed.

> **Val Sapin Variante**
> You can reach the Rifugio Bonatti by a slightly easier and much more sheltered route up the Val Sapin, useful in windy weather. Instead of following the main TMB left on to the path, continue on the vehicle track straight up the valley as far as Chapy. A path leads south, then east, climbing steeply to the pastures of Curru. The gradient increases on the switchbacks to the Col Sapin (2435 m) where you rejoin the main TMB.
>
> However in foul weather many walkers either wait it out in Courmayeur or take the bus to Planpincieux and walk up the Val Ferret. There is also a path between Courmayeur and Planpincieux.

The TMB is signed east from the Col Sapin, descending steadily into the head of the Armina Valley. The path crosses a stream and then begins to climb again, reaching another high point at the Pas entre deux Sauts (2524 m).

Descend gently past conspicuous grassy boulders into the Vallon de Malatra. The path keeps to the left (south-west) side of the valley floor.

After 1 km of relatively flat walking you'll notice prominent white buildings on the right. Ignore these, and instead follow the path steeply downhill, arriving at the Rifugio Bonatti after another 20 minutes.

3·6 Rifugio Bonatti to La Fouly

Map	panel 3 **Distance** 18 km (11 miles) **Time** 5½-6½ hours
Terrain	forest path down to the Val Ferret then a tarmac road to Arnuva; track or path to the Rifugio Elena and then a good path across the Grand Col Ferret; finally a mixture of path, track and road
Grade	steep descent followed by gradual climb to Arnuva; steep climb over the Grand Col Ferret followed by a long, steady descent
Total ascent	880 m (2885 ft) **Total descent** 1420 m (4660 ft)
Food and drink	Rifugio Bonatti, Rifugio Elena, Ferret, La Fouly
Summary	a varied stage with one high pass marking the Italian-Swiss border; views back down the Italian Val Ferret are superb

You begin with a 30-40 minute descent to reach a surfaced road in the Val Ferret. In the valley, turn right (north-east) and follow the road gently uphill for another hour to a bridge at Arnuva. There's a restaurant just off to the right.

Continue across the bridge and on to a track. Very soon the track swings left and an obvious path climbs steeply up a spur to the right – a shortcut to the Rifugio Elena that saves 10-15 minutes. The TMB is signed along the track which makes some long switchbacks before approaching the rifugio from the north.

Rifugio Bonatti
This excellent refuge is owned by Walter Bonatti, an Italian climber who made first ascents of some of the most difficult and prized routes on the peaks of the Mont Blanc massif. He is considered by many to be the greatest Alpinist of his generation.

Rifugio Bonatti, looking across Val Ferret to the Grandes Jorasses

Grand Col Ferret

The day's main ascent starts out from the left (northern) end of the rifugio and climbs steeply in a series of switchbacks. The path ascends the northern edge of a ravine and then swings away to the north around a spur. The gradient eases for the final short climb to the Grand Col Ferret (2537 m). Perched on the Italian-Swiss border, you can now rest and savour the stunning views back down the Italian Val Ferret.

The long descent into the Swiss Val Ferret begins on a wide path, dropping gently along the left (north) side of the valley headwall. After almost 3 km, the path turns abruptly south and descends steeply to a small farm building that sells refreshments.

Join a vehicle track here and follow it down through thick scrub and scattered patches of birch and aspen. Look out for unsigned paths that shortcut a couple of the longer switchbacks.

At the bottom of the descent, cross a bridge and join a surfaced road. Follow it gently downhill for 20-30 minutes through pine forest to the hamlet of Ferret.

The TMB turns left at the first buildings and crosses to the other side of the river, where it swings right and continues down the valley through pleasant forest. At a junction, there's a path going left, climbing slightly for a few minutes to reach the Gîte de la Léchere.

If you aren't heading for the gîte, keep right and follow paths and tracks to another bridge. Cross back to the right (east) bank of the river and pick up the road, reaching the village of La Fouly within 500 m.

3·7 La Fouly to Champex

Map	**panels 3 and 4**	**Distance**	**15 km (9 miles)**	**Time**	**4-5 hours**
Terrain	**pleasantly varied mixture of paths, tracks and small roads**				
Grade	**gradual descent to Issert then a steady climb to Champex**				
Total ascent	**430 m (1410 ft)**	**Total descent**	**575 m (1885 ft)**		
Food and drink	**La Fouly, Praz de Fort, Champex**				
Summary	**many walkers skip this valley walk and take the bus between La Fouly and Champex, but they miss a pleasant stage passing through quiet forest and quaint Alpine hamlets**				

Pick up the TMB along the main road through La Fouly. At the northern end of the village it turns left on to a small path that soon joins a surfaced road. Follow the road across the river and bear right on to a track.

Descend gently for 1 km and then turn left on to a narrower track. After a further 1 km this gives way to a path. There are impressive rock walls and a waterfall on your left.

You enter a pine forest, now on a track again, descending past a bridge, staying on the left (west) bank of the river. The track continues down to a sharp bend where the TMB goes left on to a path. Keep alert here: the junction may not be signed.

Climb steadily for the next 10-15 minutes, as the valley side becomes ever steeper. Cross a steep gully where a chain acts as a handrail.

Descend steeply via switchbacks to a flatter path. At a junction turn right, and continue along a curiously straight and level path built on top of a lateral moraine.

Descend, staying on the left (west) bank of the river, following a vehicle track and then a narrow surfaced road. You soon pass the first pretty chalets of Praz de Fort.

Drinking from water fountain, Les Arlaches

Continue straight ahead at a crossroads and over a bridge. A gentle descent leads to the main road and the centre of the village where there is a small café.

Turn right and cross a bridge. Almost immediately turn left on to a narrow road. Follow this down past the chalets and rustic farm buildings of Les Arlaches, keeping right where the road forks among the houses.

> **La Fouly-Champex bus**
> Several buses go daily from La Fouly to Champex via Orsières (St-Bernard Express, tel 0041 27 783 11 05). The journey takes 1½-3 hours depending on your luck with the connection in Orsières. You could instead take the bus as far as Issert and walk the rest of the way to Champex.

The road becomes a vehicle track and leads down to the main road at the hamlet of Issert. Follow the main road down through Issert before turning left on to a small road, then turn right on a track for 50 m.

Now turn left on to a path, which climbs steeply into pine forest. Follow the path past the entrance to an old mine and continue to climb steadily, past an old military installation built into a rocky outcrop. Soon you reach the main road, where you turn left and walk a short distance up into Champex.

Alpine hamlet of Praz de Fort, Val Ferret

3·8a Champex to Trient

via Bovine

Map	panel 4	**Distance**	16 km (10 miles)	**Time**	4-5 hours

Terrain tracks and small roads for the first few km; then a path, rough at times, to the Col de la Forclaz; finally tracks and paths

Grade gradual descent followed by an increasingly steep ascent to Bovine; steady descent to Trient

Total ascent 780 m (2560 ft)   **Total descent** 730 m (2395 ft)

Food and drink Champex, Bovine, Col de la Forclaz, Trient

Summary a much easier route to Trient than stage 3.8b with only one steep ascent; views down the Rhône Valley and across the Pennine and Bernese Alps are exceptional

Follow the main road north-east out of Champex past the chairlift. The more difficult TMB *variante* turns left here, see stage 3.8b.

The Bovine route stays on the main road for another 500 m and then veers left on to a track. This joins a narrow tarmac road, which leads you gently down through the chalets of Champex d'en Haut.

Watch out for the left turn on to a track. It narrows to a path, then widens out again as you climb steadily round a forested spur and into a wild, steep-sided valley.

Between Bovine and the Col de Forclaz

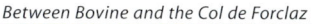

A typical waymarker

Where the track fades out a rocky path takes over, climbing steadily towards the head of the valley. The path crosses a stream as it swings to the right and begins a steep and rugged ascent to the tree line.

Now contour to the right (north-east) around a couple of spurs to reach the little chalet restaurant at Bovine. In clear weather there are great views down the Rhône Valley to the distant mountains of the Bernese Oberland. There are also great views back across the glaciers of the Grand Combin into the Pennine Alps.

Climb for a short distance on to a spur and then begin to descend on a rocky path through beautiful mixed forest. A flat meadow with some shepherd's huts breaks the descent, which then continues steadily all the way to the Col de la Forclaz (1526 m). This is crossed by the main Chamonix-Martigny road and can seem busy. On the right are the Refuge Hôtel de la Forclaz and Camping Arpille.

To reach Trient, follow signs off to the left. A path cuts a steep line down the slope before rejoining the road. Turn left and follow the road for a short distance before turning right on to a steep forest path that brings you to a track. Turn right on to this track and follow it down into the village of Trient.

Alpine orchid

3.8b Champex to Trient
via Fenêtre d'Arpette

Map	panel 4
Distance	16 km (10 miles)
Time	6-7 hours
Terrain	good paths to the Relais d'Arpette then increasingly rugged until after the Fenêtre d'Arpette; paths then improve, followed by some road into Trient
Grade	long ascent to the Fenêtre d'Arpette, very steep at the top; then a protracted steep descent
Total ascent	1185 m (3890 ft) Total descent 1385 m (4545 ft)
Food and drink	Champex, Relais d'Arpette, Chalet du Glacier, Trient
Summary	arguably the toughest stage of all, the rocky notch of the Fenêtre d'Arpette contrasts with the wide passes that have gone before; great views of the Glacier du Trient during the descent

Follow the main road north-west out of Champex but at the chairlift turn left on to a narrow road, ignoring the main TMB signs: see page 44.

Turn right on to a path that passes beneath the chairlift and enters pleasant pine forest, running along an irrigation channel. After 20-30 minutes it becomes a vehicle track. Turn left and climb past the Relais d'Arpette to flat pastures in the Val d'Arpette.

Glacier du Trient from the Fenêtre d'Arpette

Gradually the path becomes rougher and begins to climb along the right (north) side of the valley. As you rise above the tree line the path follows a stream and becomes increasingly rugged. Pause where it crosses the stream to top up your water: in late summer this will be your last chance before the other side of the pass.

After the stream, the path climbs more steeply across boulder-strewn slopes. At an unmarked junction keep right, and continue across a flat basin. Then you climb steeply across a jumble of large boulders, aiming for the prominent notch in the ridge above you. After the boulder field the path becomes braided, all options zigzagging up the very steep, narrowing slope to the Fenêtre d'Arpette (2665 m).

In contrast to the wide passes you have crossed so far, the Fenêtre d'Arpette is a rocky notch with only a few comfortable places to sit. Its austere surroundings afford fine view across the Glacier du Trient.

The first part of the descent is very steep and rocky: take care. Beyond this the path becomes more straightforward as it descends the steep open slopes above the glacier.

About 45-60 minutes from the top of the pass you reach a group of ruined buildings. The trail descends steeply across shrub-covered slopes, eventually reaching pine forest on the banks of Le Trient, after a further 45-60 minutes.

The main descent to Trient is now over. After a short walk along a stony track, you can buy meals and refreshments at the Chalet du Glacier.

From the chalet there are two options. The right-hand path contours above the Trient valley to the Col de la Forclaz, where it joins the main TMB to Trient via Bovine, see stage 3.8a. The left-hand path takes you straight to Trient via Le Peuty and is the more convenient, unless you have chosen to stay at the Col de Forclaz. For walkers with energy for another ascent, there's the option of a variante via Les Grands: see the upper panel on page 49.

To go direct to Trient, turn left and cross the river on a footbridge. Join a surfaced road and follow this for 30-40 minutes, down to the hamlet of Le Peuty. There is a simple camping area with cooking shelter and toilets, plus the very basic Refuge du Peuty (which has no warden).

To reach Trient, continue to descend along a surfaced road for a further 15 minutes. The village's small supermarket is situated past the church on the main Chamonix-Martigny road, but it keeps irregular hours.

Walkers approaching the Col de Balme

3.9 Trient to Tré le Champ

Map	panel 4
Distance	14 km (9 miles)
Time	4½–5½ hours
Terrain	road and track for the first 2 km then good paths and a short section of track for the remainder
Grade	long steady climb to the Col de Balme; downhill to Tré le Champ (except for a short climb on the Aiguillette des Posettes option)
Total ascent	960 m (3150 ft)
Total descent	830 m (2725 ft)
Food and drink	Trient, Col de Balme, Tré le Champ
Summary	a pleasant stage, returning to the Chamonix Valley with fine views; use the cable-car if you wish to skip the descent

From the centre of Trient the TMB passes the church, then turns south along a surfaced road. The road climbs past chalets to the hamlet of Le Peuty.

Bear right on to a narrow track that climbs steadily for 1 km to a forested spur. As you enter the trees, the track becomes a path and begins to climb steeply in long switchbacks. After an hour or so you emerge from the trees high above the Nant Noir. In clear weather the Refuge du Col de Balme is visible on the skyline.

> **Les Grands Variante**
> You can reach the Col de Balme without descending into Trient. Cross the bridge from the Chalet du Glacier (stage 3.8b) and take a path to the left that climbs steeply to some rock slabs. A handrail secures passage across this to the Refuge des Grands. No meals are available, but there are cooking facilities. A rocky path now contours around several spurs and ridges to the Col de Balme. Walking time from the Chalet du Glacier to the Col de Balme is about 3-3½ hours.

The path now climbs steadily across north-facing slopes for over 1 km, before joining a wide, stony track. The track winds through several switchbacks, which take you the rest of the way to the Col de Balme (2191 m).

Suddenly you can see all the way down the Chamonix Valley to the Col de Voza, almost as far as Les Houches. To the south, the snowy dome and rock pillars of the Aiguille Verte dominate the view, whilst to the south-west stand the rugged Aiguilles Rouges. The final two days of the TMB traverse along the flanks of this range.

The TMB now descends into the head of the Chamonix Valley at Tré le Champ, but there are several ways of getting down. If you're tired you can descend by cable-car to the village of Le Tour and make the short walk to either Tré le Champ or Argentière. The main TMB bears off to the right along a path that contours under the Tête de Balme before descending along a broad shoulder to the Col des Posettes (1997 m).

> **Téléphérique Le Tour**
> Some walkers take the chair lift and cable-car from the Col de Balme down to Le Tour (tel 0033 4 50 54 00 58), alighting a short walk from either Tré le Champ or Argentière. To reach the top station, follow the obvious path to the left from the Col de Balme for about 10 minutes. In 2004, the one-way fare was €10.

It then climbs on to and along the crest of the Aiguillette des Posettes (2201 m), which affords 2-3 km of fine ridge walking with added views across Vallorcine.

Trient 1297 — 2191 Col de Balme — 2201 L'Aiguillette des Posettes — *Tré le Champ* 1417

If you don't want to make the extra 200 m of ascent, keep straight ahead at the Col des Posettes and contour around the slopes to the left of the Aiguillette des Posettes. The views are also good from this path, but are limited to the Chamonix Valley and the Mont Blanc massif. As you begin to descend, keep right at a trail junction, following signs for Tré le Champ.

The main TMB and side-path reunite near the tree line. A single path now winds steeply through pine forest to reach the main Chamonix-Martigny road, just slightly uphill from Tré le Champ.

Turn left and follow a track that runs alongside the road. A five-minute walk brings you to the Gîte La Boerne, which is about all there is of Tré le Champ. There's much more choice of services and accommodation in the small town of Argentière, a further 20-30 minutes walk down the valley along pleasant tracks and paths. If you overnight in Argentière, you can rejoin the TMB without returning to Tré le Champ. A path leads from the north end of the town directly to La Tête aux Vents (see stage 3.10).

Refuge de la Balme

3·10 Tré le Champ to La Flégère

Map	panel 4 **Distance** 7 km (4½ miles) **Time** 3-4 hours	
Terrain	good but sometimes rocky paths throughout; exciting section of iron ladders above Tré le Champ	
Grade	a steady ascent at first, but very steep from the ladders to La Tête aux Vents; thereafter a gradual descent to La Flégère	
Total ascent	790 m (2590 ft) **Total descent** 330 m (1085 ft)	
Food and drink	Tré le Champ, La Flégère	
Summary	after the ladder system, a superb walk along the Grand Balcon Sud, with great views of Mont Blanc and a good chance of seeing ibex; in good weather, divert to Lac Blanc for even better views	

Tré le Champ 1417 — Tête aux Vents 2130 — La Flégère 1877

Chamonix Aiguilles and one of the Lacs des Cheserys

Although there is almost 800 m of ascent to La Flégère, this stage takes only half a day. If you have time and energy to spare, you could take in the side trip to Lac Blanc, where you can also spend the night. In fine weather the extra effort to reach Lac Blanc will be well rewarded by some of the best views of the Mont Blanc massif to be found anywhere on the TMB.

The trail to La Flégère leaves the Chamonix-Martigny road just 100 m downhill from where the trail came in from the Col de Balme at the end of the previous stage. Climb past a slab of rock and through mixed forest and then turn right at a trail junction.

> **Lac Blanc**
> From La Tête aux Vents follow a path uphill, signed for Lac Blanc and Lacs des Cheserys. A steady ascent leads round the back of a small outcrop, down and around one of the Lacs des Cheserys. The path then climbs steeply, often crossing a bank of snow, to reach a small ladder.
> Above the ladder, until late summer, expect more snow on the final steep pull to the Refuge du Lac Blanc, set on a rock rib just above the lake. To descend from Lac Blanc follow the path signed for La Flégère, where you rejoin the main TMB.

The path climbs steadily past the tree line towards the bottom of a band of imposing cliffs. There is a dramatic free-standing pinnacle of rock on the left, the Aiguillette d'Argentière.

An exciting series of iron ladders and wooden walkways forces a passage up a steep gully. Take care here, especially if your pack is heavy. The first ladder is the longest and steepest, so if you can cope with this then you should be fine with the rest.

Alpenrose and the Argentière Glacier from the Balcon Sud

Tackling ladders above Argentière

If you find the first ladder too difficult, return to the last junction and follow signs towards Argentière, where you'll meet a more straightforward path to La Flégère. (If you suffer from vertigo, you might not want to go near the ladders.) Or, you can walk or take the bus from Tré le Champ to the Col des Montets and follow the steep but straightforward path from there to La Tête aux Vents.

From the top of the ladders, the trail continues climbing steeply to a large cairn and trail junction at La Tête aux Vents, about two hours from Tré le Champ. La Flégère is signed to the left along the Grand Balcon Sud, whilst the trail to Lac Blanc continues uphill from the cairn. The path to La Flégère descends past several stone buildings and crosses a stream beneath a band of cliffs. It is not uncommon to see chamois or ibex in this area, particularly early in the morning or later in the evening when the trail is quieter. It then goes round a rocky spur before descending steadily to the cable-car station and the Refuge de la Flégère.

If for any reason you need to descend to Chamonix, use the Téléphérique de la Flégère which takes you to the village of Les Praz de Chamonix just a few kilometres north-east of Chamonix itself.

The Chamonix Aiguilles

3.11 La Flégère to Les Houches

Refuge Bellachat, Chamonix Valley

The final stage to Les Houches is fittingly spectacular, but very demanding if you hike all the way from the top of Le Brévent down to Les Houches, a descent of over 1500 m (5000 ft). Some walkers instead let the Téléphérique du Brévent sweep them directly to the cafés and restaurants of Chamonix; see the panel on page 57.

Map	panel 1
Distance	18 km (11 miles) **Time** 5½–6½ hours
Terrain	rocky paths and tracks to the summit of Le Brévent; steep but good path for the descent to Les Houches
Grade	steady climb to Le Brévent with some steeper sections; very long descent to the finish
Total ascent	770 m (2525 ft) **Total descent** 1615 m (5300 ft)
Food and drink	La Flégère, Planpraz, Le Brévent, Refuge de Bellachat, Les Houches
Summary	sublime views of Mont Blanc and the Chamonix Aiguilles; use the cable-car to avoid the gruelling descent to Les Houches

From La Flégère follow signs for Planpraz and contour beneath a chairlift and around a steep-sided basin. After about an hour you cross a vehicle track and climb up under another chairlift. A steady ascent now takes you to the rather charmless cable-car station and restaurant at Planpraz.

Now follow signs for the Col du Brévent. A rocky path climbs steeply on switchbacks to reach a large cairn marking the col, about 45 minutes after Planpraz. There are great views from here over the mountain ranges to the north-west.

Bear left from the col, along a ridge and then through a boulder field. Cross a rocky shoulder with the help of some handrails. At its top join a vehicle track where you turn right.

After another 10 minutes of steady ascent, you reach the summit restaurant and cable-car station on Le Brévent (2525 m). The views of Mont Blanc are good, but the cable-car carries many sightseeing tourists and the crowds can seem overpowering.

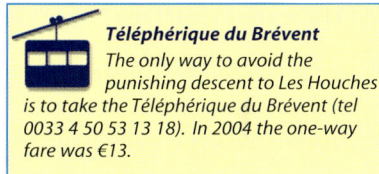

Téléphérique du Brévent
The only way to avoid the punishing descent to Les Houches is to take the Téléphérique du Brévent (tel 0033 4 50 53 13 18). In 2004 the one-way fare was €13.

Return back along the track for 100 m and turn left on to a stony, zigzagging path. The descent from Le Brévent is initially quite steep, but soon eases as you progress along a gentle shoulder dotted with small tarns and pools.

After an hour you reach the Refuge de Bellachat. Here the route splits: the trail on the left descends directly to Chamonix, whereas the TMB follows the right-hand trail to Les Houches.

After a section of steep switchbacks, the descent becomes steadier through the pine forest. The gradient eases further as you pass the northern boundary of the Merlet Animal Park, but drops steeply again past the statue of Le Christ Roi.

The path now swings to the south for the final section of the descent, bringing you down to a surfaced road. Turn right and follow the road to Les Houches SNCF station. Cross a bridge over L'Arve river and follow the road for the final stretch into the centre of Les Houches.

Here you can take off your pack for the final time and perhaps enjoy a well-earned drink to celebrate the end of one of the world's great walks.

Les Houches

4 Reference

Accommodation list

This list shows selected accommodation including all refuges outside towns and villages. Book your place in a refuge ahead, especially during July and August. For accommodation in towns, try the local tourist information office first.

Name	Tel	Location
Refuge Michel Fagot	+33 4 50 54 42 28	Les Houches
Gîte du Champel	+33 4 50 47 77 55	middle stage 1a
Refuge du Fioux	+33 4 50 93 52 43	middle stage 1a
Auberge Bionnassay	+33 4 50 93 45 23	middle stage 1a
Refuge de Miage	+33 4 50 93 41 03	middle stage 1b
Refuge du Truc	+33 4 50 93 12 48	middle stage 1b
Chalet du CAF	+33 4 50 47 00 88	Les Contamines
Camping Le Pontet	+33 4 50 47 04 04	Les Contamines
Gîte d'étape Pontet	+33 4 50 47 04 04	Les Contamines
Refuge Nant Borrand	+33 4 50 47 05 45	middle stage 2a
Refuge de la Balme	+33 4 50 47 03 54	middle stage 2a
Refuge de la Col du Croix du Bonhomme	+33 4 79 07 05 28	middle stage 2a/start 2b
Auberge de la Nova	+33 4 79 07 05 28	Les Chapieux
Refuge des Mottets	+33 4 79 07 01 70	end stage 2b/middle 3
Rifugio Elisabetta	+39 1 65 84 40 80	end stage 3
Rifugio Maison Vieille	+39 1 65 80 93 99	middle stage 4
Rifugio Monte Bianco	+39 1 65 86 90 97	middle stage 4
Hotel Venezia	+39 1 65 84 24 61	Courmayeur
Rifugio Bertone	+39 1 65 84 46 12	middle stage 5
Rifugio Bonatti	+39 1 65 86 90 55	end stage 5
Rifugio Elena	+39 1 65 84 46 88	middle stage 6
Hotel Col de Fenêtre	+41 27 783 11 88	Ferret
Gîte de la Léchere	+41 27 783 30 64	La Fouly

Chalet le Dolent	+41 27 783 18 63	La Fouly
Hotel de Glaciers	+41 27 783 11 71	La Fouly
Camping des Glaciers	+41 27 783 17 35	La Fouly
Camping Rocailles	+41 27 783 19 79	Champex
Chalet Plein Air	+41 27 783 23 50	Champex
Auberge Bon Abri	+41 27 783 14 23	Champex
Chalet du CAS	+41 27 783 11 61	Champex
Refuge Hôtel de la Forclaz	+41 27 722 26 88	end stage 8a
Camping Arpille	+41 27 722 26 88	end stage 8a
Refuge Les Grands	+41 26 658 13 23	alternative route
Relais d'Arpette	+41 27 783 12 21	start stage 8b
Refuge du Peuty	+41 27 722 09 38	Le Peuty
Gîte la Gardienne	+41 27 722 12 40	Trient
Café Gîte Moret	+41 27 722 27 07	Trient
Refuge du Col de Balme	+33 4 50 54 02 33	middle stage 9
Gîte la Boerne	+33 4 50 54 05 14	Tré le Champ
Refuge de la Flégère	+33 4 50 53 06 13	end stage 10
Refuge du Lac Blanc	+33 4 50 53 49 14	alternative route
Refuge de Bellachat	+33 4 50 53 46 99	middle stage 11
Gîte La Montagne	+33 4 50 53 11 60	Chamonix
Le Chamoniard Volant	+33 4 50 53 14 09	Chamonix
Camping La Mer de Glace	+33 4 50 53 08 63	Les Praz de Chamonix

Traditional alpine refuge, Bionnassay

Books

McCormack, G et al *Walking in the Alps* Lonely Planet, 384pp, 2004, 1-74059-395-2

Comprehensive and reliable; gives additional coverage of day walks in the Mont Blanc area and useful information on Alpine walking.

Stefano Ardito *Mont Blanc: Discovery and Conquest of the Giant of the Alps* White Star Editions, 228pp, 2001, 8-88095-102-5

Detailed history of mountaineering on Mont Blanc, from Balmat and Paccard to the modern era, well illustrated throughout.

Walt Unsworth *Savage Snows: The Story of Mont Blanc* Hodder & Stoughton, 192pp, 1986, 0-34039-777-2

Maps

Rando Editions publishes the *Pays du Mont Blanc* map covering the entire route at 1:50,000. The IGN (Institut Géographique National) has sheets 3530ET *Samoëns Haut-Giffre*, 3531ET *St-Gervais-les-Bains Massif du Mont Blanc*, and 3630OT *Chamonix Massif du Mont Blanc* that all but cover the route at 1:25,000. For part of stage 5 you need the IGC (Instituto Geografico Centrale) map No 107 *Monte Bianco*.

The IGN and Rando Editions maps are widely available from outdoor shops in Chamonix, and the IGC map is available in Courmayeur, or try buying online from specialist map sellers.

Useful contacts

Office de Haute Montagne
www.ohm-chamonix.com
tel +33 4 50 53 22 08
Provides a useful accommodation list, advice on trail conditions and weather forecasts.

Chamonix Tourist Information
www.chamonix.com
+33 4 50 53 00 24
Efficient office provides useful details on accommodation, public transport, refuges and cable-cars

British Mountaineering Council
www.thebmc.co.uk
tel 0870 010 4878
BMC members receive discounts on accommodation in some refuges.

Guided climbs and tour operators

Trekking in the Alps
www.trekkinginthealps.com
tel +33 4 50 54 62 09

La Compagnie des Guides de Chamonix
www.chamonix-guides.com
tel +33 4 50 53 00 88

Great Walks of the World
www.greatwalks.net
tel 01935 810 820

Weather forecasts

Three-day mountain weather forecasts for the Mont Blanc area are given in English at **http://meteo.chamonix.com**. Recorded English forecasts can be heard by phoning +33 8 92 70 03 30. Detailed forecasts are also posted daily outside information offices and outdoor shops in Chamonix and Courmayeur.

Transport

British Airways
www.ba.com
tel 0845 084 4444 (UK)

Ryanair **www.ryanair.com**
tel 0871 246 0000 (UK)

Easyjet **www.easyjet.com**
tel 0871 7 500 100 (UK)

Geneva Airport
www.gva.ch
tel +41 22 717 71 11

Turin Airport
www.turin-airport.com
tel +39 115 67 63 73

Chamonix SNCF station
www.sncf.com
tel +33 4 50 53 00 44

SAT **www.sat-montblanc.com**
tel +33 4 50 53 01 15

Notes for novices

If you're new to long-distance walking, please see our notes on preparation and gear. Find them on our website at **www.rucsacs.com**, or send a suitably stamped addressed envelope marked 'Notes for novices' to: Rucksack Readers, Landrick Lodge, Dunblane, FK15 0HY, UK.

Acknowledgements

The author wishes to thank Helen Fairbairn, Michael Thornhill, Colin Killeen, Kerril Thornhill, Stuart Fairbairn and Pól O'Seasnain.

Photo credits

Gareth McCormack (all images)
www.garethmccormack.com

Rucksack Readers

Uniform with this volume are *Explore the Inca Trail (2002)*, *Explore the Great Wall (2003)* and *Explore Mount Kilimanjaro (3rd ed, 2005)*. We also publish eight books on classic long-distance walks in Scotland and Ireland.

For more details, or to order online, please visit **www.rucsacs.com** or telephone +44/0 1786 824 696.

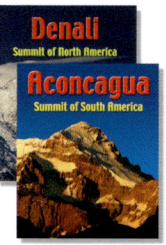 ISBN 1-898481-51-2

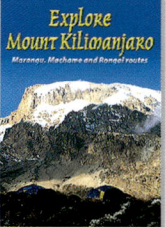 ISBN 1-898481-23-7

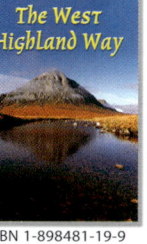 ISBN 1-898481-19-9

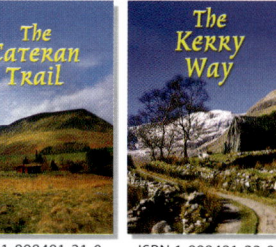 ISBN 1-898481-21-0

ISBN 1-898481-22-9

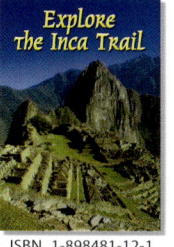 ISBN 1-898481-12-1

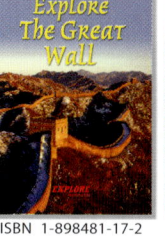 ISBN 1-898481-17-2

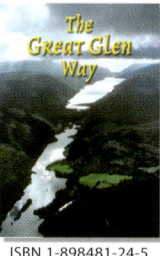 ISBN 1-898481-24-5

ISBN 1-898481-13-X

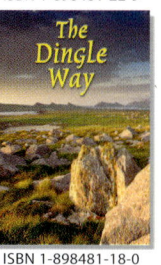 ISBN 1-898481-18-0

Index

A
acclimatisation 16
accommodation 11, 59-60
Alpine gentian 19, 20
Alpine orchid 45
altitude sickness 16
altitude profile 6

B
baggage 6, 12, 13
Beaufort cheese 31
Bonatti, Walter 14, 40

C
cable-cars (téléphériques) 5, 8, 9, 17, 22, 23, 36, 49, 57, 61
camping 12, 13, 17
Chamonix 4, 9, 10, 55, 56, 57, 61
Champex 9, 42-3, 44
Col des Fours variante 30-1
Col de Tricot variante 25-6
Courmayeur 4, 6, 10, 34, 36-7, 38, 61

D
dehydration 8

F
Fenêtre d'Arpette variante 46-7
food and drink 6-7, 8, 12

G
geology 18
glaciers 18, 34, 35, 37, 45
gîtes 5, 11, 12, 59-60
guides 15, 61

H
habitats and wildlife 19-21
history of Mont Blanc 14

I
ibex 20, 51
iodine 8, 13

L
La Flégère 4, 20, 51-5, 56
La Fouly 9, 40-1, 42
Lac Blanc 4, 51, 52, 55

Lacs des Cheserys 4, 52 ??
Les Contamines 6, 22, 24, 27
Les Chapieux 6, 27, 32
Les Grands variante 49
Les Houches 6, 9, 10, 22, 49, 56-8

M
marmot 21
mobile phones 8, 13
Mont Blanc, ascent of 15-17, 61
moss campion 19

N
Notes for novices 8, 62

P
pace and elevation 6
packing and checklist 12, 13
planning and preparation 5-13

R
refuges, alpine 6, 8, 9, 11-12, 17, 27, 29, 30, 59-60, 61
Refuge des Mottets 30, 31, 32
Rifugio Bonatti 38, 39, 40
Rifugio Elizabetta 12, 32, 33, 34

S
safety, personal 8, 17

T
time of year, best 5, 8, 11
tour operators 6, 16, 61
travel and transport 10, 62
Tré le Champ 9, 20, 48-50, 51
Trient 44-5, 48

V
Val Ferret 7, 40, 41 ??
Val Sapin variante 39
variantes 7, 25-6, 30-1, 38, 39, 44, 46-7, 49

W
water, drinking 8
weather and forecasts 8, 16, 61
wildlife and habitats 19-21